The Masca Gorge

The Masca Gorge on Tenerife, Spain in the Canary Islands

Other Books by Pia Lord

Tenerife Canary Islands, Spain
El Teide Tenerife Volcano Canary Islands Spain
Return to the Sea: A Guide to Open Water Swimming in Mallorca, Spain

Catskills in the Rocky Mountains
Cato the Caterpillar
The Night the Moon Went Out
The Day the Sun Went Out
Time and Travel
The Upper Limit
Let's Take a Trip in our Spaceship
Aggity Biggity Ciggity

Rhapsody
The ABC Collection
Harvest While the Orchard is Aplenty
The Adventures of M.M. Music Mouse
Just Pia!

Dedication

This little book is written to inspire nature lovers everywhere to hike the Masca Gorge and discover its richness in its pristine state. It is important to know that there are still corners of the Earth relatively untouched by modern machinery. These areas should remain so and furthermore, we should let the Earth reclaim many more spaces so that she can heal our planet. Preserving our Earth is one of the greatest challenges of the 21st century with our present habits and lifestyles in this vehicle driven consumer economy.

Acknowledgements

Thank you to my family and friends for providing moral support in times that I need it and being supportive of my interests. I believe the Masca Gorge is one of important regions of the Earth that should be highlighted for its untouched beauty. The more we teach teachers to teach children to value the Earth, beauty and the pristine nature of our planet, the more children will grow up caring for what we have and trying to keep it in its natural state. They will choose beauty, clean air and healthy living over built up, smogged up, pollution ridden waters and lands any day. They must learn what the Earth is really like and was like in the heavily populated areas prior to the industrial revolution,and the machine and building ages. Thank you also to teachers who already teach environmental conservation. I also would like to thank the guides at SwimTrek for leading me to this gorge. And last but not least I would like to thank my husband and son for manning the fort while I am gone.

Tenerife's Masca Gorge

Canary Islands, Spain

by
Pia Lord

Sunlight and shadow on cliffs approaching the Masca Gorge. La Gomera in the Atlantic Ocean in the distance. The town of Masca is sunlit in the foreground.

Massive lava formations appear perched and carefully balanced.

Caves in the varied texture of the lava flows and the lichen growth in the Masca Gorge.

Layered basalt rock lava flows and traces of waterfalls in the Masca Gorge.

A cool walk down the gorge with cliffs partially in shadows

High cliffs of the gorge illuminated by rays of sunlight peering through.

Cairns sitting upon lichen covered volcanic boulders.

Lava flows as rising cliffs about 800-1000 meters high.

Wild grasses growing in Masca Gorge surrounding the lava caves.

A small pond with crystal clear waters.

Hikers traverse undulating rock formations, negotiating the ascent in the Masca Gorge.

Trees, bushes, lichen and mosses surround a dried up waterfall.

Palm trees growing in sandy soil in the depths of the gorge.

Wild giant aloe plants and prickly pear cactus grace the slopes of the gorge.

Fields of giant aloe on the steep slopes of the gorge make for interesting climbing for those in good shape.

The two to three-hour hike from the town of Masca is about an 8k journey on foot descending into the gorge.

The prickly pear cactus bears fruit in abundance on these hillsides.

Shadows rise and grow in the afternoon creating dramatic contrasts on the cliffs.

Survival of the most adaptable in this gorge on the island of Tenerife.

Pointed cliff tops become visible upon approach to the sea.

Playa Masca, the beach at the end of the Masca Gorge hike, is filled with lava boulders.

Due to boat traffic, there is a thin film of oil floating on the surface of the near pristine waters of the North Atlantic Ocean. La Gomera, Canary Island is in the background.

Activities and Research Projects

A. Fun Geology Activities to Think About and Do

1. Go outside and find interesting rocks. Bring them in and start a collection. Get a notebook, and write down each rock name, take a picture, add characteristics. Over the course of several months or years, develop an interesting collection of rocks that you find in your travels.

2. With the permission of adults, parents or teachers, take a field trip to a geologically interesting location. This can be a gorge, a canyon, a plain, a volcano, a crater, a caldera, a mesa, Arches National Park, mountains nearby where you live, the water's edge, coastal areas etc. Look for rocks and evidence of geological layers in the earth. Write up an interesting paper with illustrations describing the location and what sort of geological layers you found.

3. Make a shoebox diorama showing an area of the earth that you found to be geologically interesting. Use colors and labels to add detail for understanding of the structure.

B. Different Types of Rocks Found in Gorges.
Look up famous gorges and study the rocks that are on them. List the gorges along with the rock types. Make lists of these types of rocks.
Try to group or classify gorges and rocks.

_____

_____

_____

_____

_____

_____

C. The Major Gorges on Earth

Do more research on the internet and in books. Find the 5 largest Gorges on Earth.
Determine each location. Is the gorge near a large body of water as is the Masca
Gorge? What is the height from sea level to top of the gorge? What is the difference
between a canyon and a gorge? List the information below.

1._____

2._____

3._____

4._____

5._____

Look up more information on gorges/canyons. List ten more gorges/canyons and the
regions of the Earth in which they are found.

6._____

7._____

8._____

9._____

10._____

11._____

12._____

13._____

14._____

15._____

D. Map out the locations of all these gorges/canyons. For each gorge, chart information which includes the name, the height, the continent, the country, the tectonic plate and the type of formation. Determine whether gorges serve any significant purpose to the earth. Are they byproducts of other Earth processes or do they form for a particular reason? See if you can find patterns among the formations of gorges to understand how the Earth is acting and reacting. Use GIS software or Google maps, if possible, when locating and mapping the gorges. Use Excel to create your chart. Create legends for your maps. Discuss your thoughts and findings in small groups or teams, displaying your detailed maps to aid in communicating your information.

1. Make a chart of information on gorges
2. Create a map of gorges
3. Pick one gorge and write up an analysis of how it formed. Use pictures and pattern analysis to explain your ideas.
4. Global scale gorges: On a larger scale think about patterns of gorge formation. Come up with your own ideas on why gorges form. Do they have a purpose in forming?
5. Discussion Groups

Notes

_____

_____

_____

_____

Credits

Pia Lord hiking in the Masca Gorge.

About the Author

Pia Lord holds an M.S. in Space Science and attended undergraduate college at Columbia University, Barnard College in NYC. She founded the Pia Lord Company, an adventure travel company and has written 16 books in various genres on topics including space science, music and emotions, environmental science, planetary physics, orbital debris, biological life cycles and sports skiing stories. She enjoys travelling and writing. Her interests include the Earth and its oceans, reading, open water swimming, skiing, playing the piano, the flute and singing first soprano in church choir. She lives with her husband and son in New Jersey, United States of America. Her website is **pialord.com**. Her books are found at the **Pia Lord Author Page at Amazon**.

Rainbows over lava flows on Tenerife. The Canary Islands, Spain are located at 28° North Latitude and 16° West Longitude in the North Atlantic Ocean off the coast of Morocco and the Sahara Desert in northwestern Africa.

www.ingramcontent.com/pod-product-compliance
Lightning Source LLC
Chambersburg PA
CBHW041257180526
45172CB00003B/880